EU Civilian Crisis Management

The Record So Far

Christopher S. Chivvis

Prepared for the Office of the Secretary of Defense
Approved for public release; distribution unlimited

The research described in this report was prepared for the Office of the Secretary of Defense (OSD). The research was conducted in the RAND National Defense Research Institute, a federally funded research and development center sponsored by OSD, the Joint Staff, the Unified Combatant Commands, the Department of the Navy, the Marine Corps, the defense agencies, and the defense Intelligence Community under Contract W74V8H-06-C-0002.

Library of Congress Cataloging-in-Publication Data is available for this publication.

ISBN 978-0-8330-4919-3

The RAND Corporation is a nonprofit research organization providing objective analysis and effective solutions that address the challenges facing the public and private sectors around the world. RAND's publications do not necessarily reflect the opinions of its research clients and sponsors.

RAND® is a registered trademark.

Cover image: EULEX/Enisa Kasemi

© Copyright 2010 RAND Corporation

Permission is given to duplicate this document for personal use only, as long as it is unaltered and complete. Copies may not be duplicated for commercial purposes. Unauthorized posting of RAND documents to a non-RAND Web site is prohibited. RAND documents are protected under copyright law. For information on reprint and linking permissions, please visit the RAND permissions page (http://www.rand.org/publications/permissions.html).

Published 2010 by the RAND Corporation
1776 Main Street, P.O. Box 2138, Santa Monica, CA 90407-2138
1200 South Hayes Street, Arlington, VA 22202-5050
4570 Fifth Avenue, Suite 600, Pittsburgh, PA 15213-2665
RAND URL: http://www.rand.org/
To order RAND documents or to obtain additional information, contact
Distribution Services: Telephone: (310) 451-7002;
Fax: (310) 451-6915; Email: order@rand.org

Preface

Since 2000, the European Union has been developing civilian capabilities for use in civilian missions, including postconflict and other environments. The EU has deployed civilian experts in a variety of capacities to Iraq, Afghanistan, and Kosovo, as well as other countries in Europe, the Middle East, Africa, and Asia. As the United States assesses and develops its own civilian capabilities, it will be important to understand what the EU is capable of doing in this area. This report looks at the record of EU civilian operations so far, drawing conclusions both for the United States and Europe.

This research was sponsored by the Office of the Secretary of Defense and conducted within the International Security and Defense Policy Center of the RAND National Defense Research Institute, a federally funded research and development center sponsored by the Office of the Secretary of Defense, the Joint Staff, the Unified Combatant Commands, the Navy, the Marine Corps, the defense agencies, and the defense Intelligence Community.

For more information on RAND's International Security and Defense Policy Center, contact the Director, James Dobbins. He can be reached by email at dobbins@rand.org; by phone at 703-413-1100; or by mail at the RAND Corporation, 1200 South Hayes St., Arlington, VA 22202. More information about RAND is available at www.rand.org.

Contents

Figures and Tables

Figures

Tables

Summary

Since the end of the Cold War, the value of civilians in postconflict stabilization has become increasingly clear. As a result, beginning in 2003, the European Union began deploying civilian missions under the auspices of the European Security and Defence Policy (ESDP). In contrast with the EU's military missions, however, these civilian missions have received little attention. This report provides an early assessment of the EU's civilian record, examines the challenges ahead, and outlines the main policy implications for the United States and Europe.

The EU has deployed civilians in several capacities and a variety of environments, ranging from benign to hostile. At the same time, the EU has continually reformed and worked to rationalize the relevant institutions in Brussels to improve its civilian record. In general, however, the EU's civilian missions have been relatively small scale and have not had a major impact on security challenges of significance to the United States. ESDP civilian work has in most cases been ancillary to larger, ongoing nation-building work.

Nevertheless, the EU is apt to do more in the future, and the record shows that the EU has managed to make valuable civilian contributions in conflict and postconflict environments, especially when they are close to Europe. Although the EU has often fallen short of its own goals, especially when it comes to staffing, and has encountered frequent logistical and planning problems, the general trend is positive. Provided that European states continue to invest in developing civilian capabilities, the EU can be expected to make a growing contribution in years ahead.

The EU's expertise in the rule of law is particularly welcome, and further development of EU capabilities in this area should be strongly encouraged. European police and legal advisers are developing the capabilities and experience necessary to bolster the rule of law in states emerging from conflict. To be more effective in building the rule of law in the future, Europe will need to expand its capabilities for executive policing and develop the ability to conduct higher-volume police training. Most of all, however, it will need to improve its record in meeting its own staffing targets.

The EU's two most important missions have both operated alongside NATO: the integrated rule of law mission in Kosovo and the EU police-training mission in Afghanistan. The Afghanistan mission has been plagued with problems and continues to underperform, despite some recent improvements. By contrast, the Kosovo mission has been successful and clearly had a positive impact on the ground, despite the several challenges it faces. There are reasons to hope that Kosovo, not Afghanistan, will be the future model.

In general, future EU contributions can be expected to be greater in regions closer to Europe—not only because European states tend to see a greater interest in these regions, but also because proximity facilitates the recruitment of civilian staff.

Main Policy Recommendations

For the United States

- Recognize that EU civilian capabilities remain limited, but are poised to become more significant in supporting allied security objectives in the future.
- Continue to support the EU's efforts to build civilian capabilities, including by taking a benevolent attitude toward the EU's Civilian Planning and Conduct Capability (CPCC) and by providing staff and logistical support when appropriate.
- Continue to work to fix the EU-NATO working relationship. While adept efforts on the ground have avoided much of the pos-

sible damage caused by the Turkey-Cyprus dispute, the problem still needs resolution if EU civilian work is to contribute to broader allied goals and operations in the future.

For the European Union

- The EU needs to focus on overcoming its staffing shortfalls. The three most promising directions for doing so are (1) further increasing EU funding for civilian missions, (2) considering more widespread use of contractors, and (3) developing a civilian reserve corps, preferably with a standing pool of staff trained and ready for deployment within 48 hours' notice. The latter would be the most ambitious option, but should not be beyond Europe's reach.
- Establish a European facility to review lessons learned from civilian missions, in order to obtain the full benefit of conducting such missions under the EU.
- Most immediately, ensure that the missions in Afghanistan and Kosovo are successful. The mission in Afghanistan is particularly at risk for lack of resources. The EU must recognize what is on the line in Afghanistan for ESDP; significantly expand the Afghan mission, even beyond the current authorized levels; and give high priority to ensure full staffing. On Kosovo, the EU must stay its course and ensure that Serb machinations north of the Iber River do not derail this flagship effort.

Acknowledgments

Thanks to Jolyon Howorth, Robert Hunter, and Giovanni Grevi for very helpful comments on an earlier version of this report. Thanks also to Joya Laha for valuable research assistance, especially on Afghanistan.

Abbreviations

CIVCOM	Committee for the Civilian Aspects of Crisis Management
CPCC	Civilian Planning and Conduct Capability
CSTC-A	Combined Security Transition Command–Afghanistan
DRC	Democratic Republic of Congo
ESDP	European Security and Defence Policy
EUJUST LEX/Iraq	European Union Integrated Rule of Law Mission for Iraq
EULEX Kosovo	European Union Rule of Law Mission in Kosovo
EUMS	European Union Military Staff
EUPOL Afghanistan	European Union Police Mission in Afghanistan
GPPO	German Police Project Office
ICO	International Civilian Office
IPCB	International Police Coordination Board
ISAF	International Security Assistance Force
KFOR	NATO Kosovo Force
NTM-A	NATO Training Mission–Afghanistan

NTM-I	NATO Training Mission–Iraq
PRT	Provincial Reconstruction Team
PSC	Political and Security Committee
SG/HR	European Union Secretary General/High Representative for Common Foreign and Security Policy
SSR	Security Sector Reform
UNMIK	United Nations Mission in Kosovo

CHAPTER ONE

Introduction

Over the past few years the U.S. policy community has moved toward a consensus on the critical importance of civilian or "civilian-military" work to the success of the military operations in which U.S. armed forces are engaged.[1] Developing effective U.S. civilian capabilities, however, requires understanding the capabilities of other actors and international organizations in particular, not only because these organizations have much to contribute, but also because U.S. staff will more often than not be expected to coordinate with them. Among these international organizations, the European Union, with the United Nations, is one of the most prominent. This report offers a preliminary look at what the EU has done in the civilian field, with an eye to improving the planning and coordination of U.S.-EU efforts, within NATO and beyond.

Many experts believe that the European Union has a special role to play in civilian work around the world. Some have argued that the multidimensional nature of the EU makes it inherently better suited for civilian work when compared with NATO, whose mission has historically been military in nature. Others argue that, even if NATO were able to undertake civilian work in crisis zones, the EU would still have a comparative advantage, given that it will never develop military capabilities on par with NATO. From this perspective, encouraging the EU's civilian capabilities will also help ensure complementar-

[1] Representative studies include Hunter et al., 2008; Barton et al., forthcoming; Cohen and Unger, 2008; and Thomson and Serwer, 2007.

ity between the two organizations and thereby, perhaps, better overall transatlantic security cooperation.

But how well suited is the EU, in reality, to civilian work? What does the empirical record show? What does the future look like? This report takes up these and related issues. Intended primarily for a U.S. policy audience, it offers a general overview and assessment of the EU's civilian operations to date, as well as a more in-depth look at the two missions in which the EU has worked alongside NATO: Afghanistan and Kosovo. These two missions are also the EU's most ambitious civilian missions and are useful for comparison, since one is widely viewed as underperforming (the EU Police Mission in Afghanistan [EUPOL Afghanistan]), while the other could still succeed (the EU Rule of Law Mission in Kosovo [EULEX Kosovo]). The report concludes with an assessment of the EU's work to date and the future outlook.

Outlook

In general, the record shows that the EU has managed to make valuable civilian contributions in conflict and postconflict environments, especially in Europe's vicinity. The main value of the EU itself has been in aggregating and coordinating European national resources. The EU does have advantages over its member states, although its natural advantages over NATO in this area tend to be much exaggerated. Although the EU has repeatedly fallen short of its own goals, especially when it comes to staffing, and has encountered frequent logistical and planning problems, the general trend is positive. Provided that the EU member states continue to invest in developing civilian capabilities, the EU can be expected to make a growing contribution.

Hence, while it is very unlikely that the EU will be able to fulfill the majority of allied needs for civilian postconflict work, the United States should continue to encourage the development of EU civilian capabilities and ensure maximum coordination in the field and in civilian force planning. That said, it would be detrimental if the EU were to focus exclusively on civilian capabilities at the expense of conventional forces. European potential in civilian work is not sufficient to warrant any such division of labor. Not only would a further decline in European military capabilities be harmful to NATO, it would also undermine European security, since most security tasks today require a combination of civilian and military power.

More specifically, the EU's focus and growing expertise in the rule of law should be welcomed, and further development of EU capabilities in this area strongly encouraged. European contributions in mentoring and advising police in postconflict situations have proven successful in some situations and may prove more successful in the future. European executive policing has also been beneficial, although mentoring and advising are the EU's preferred mode of operation. Similarly, European legal advisers are developing the capabilities and experience necessary to balance the competing tasks of bolstering the rule of law in states emerging from conflict. To be more effective in building the rule of law in the future, Europe will need to expand its capabilities for executive policing and develop capabilities for higher-volume police training. Most of all, it will need to offer enhanced incentives for staff to deploy in postconflict environments.

In general, future EU contributions can be expected to be greater in regions closer to Europe—not only because European states tend to see a greater interest in these regions, but also because proximity facilitates the recruitment of civilian staff.

CHAPTER TWO

The European Union's Civilian-Military Capabilities

The EU's Civilian Aspirations

Developing civilian capability has long been viewed as crucial to the success of the broader European Security and Defence Policy (ESDP). ESDP was launched in 1999, at a summit in St. Malo, France, and has developed into a mechanism by which members of the European Union can take joint military action to respond to crises with combined military and civilian power. Many of the driving ideas behind ESDP are rooted in the experience of the Balkan crises of the 1990s and the belief that the European Union was better equipped than NATO to handle the postconflict reconstruction phases of these conflicts and their civilian dimension in particular. The desire to develop capabilities for deploying civilians in crisis and postconflict situations has therefore been part of ESDP from the start, though EU civilian work has received less attention than EU military missions.

EU civilian missions are similar to EU military missions in that staff come from EU member states and costs are shared between member states and the EU budget. The EU, like NATO, has few assets of its own, though it has developed some planning capacity for civilian missions. EU civilian missions are under the authority of the intergovernmental EU Council rather than the supranational European Commission. The main added value of the EU, therefore, is to aggregate the resources and coordinate the efforts of Europe's national states and direct them toward common purposes under an EU flag.

EU member states began to commit to the establishment of EU civilian capabilities at the Feira European Council meeting in 2000 (European Council, 2000). At the time, they focused on four areas:

- police
- civil administration
- rule of law
- civil protection.

Two further areas were added in 2004:

- monitoring
- supporting EU Special Representatives.

More recently, security sector reform (SSR) has also been added. A "civilian headline goal" was also established in 2004, setting out the aim of building the capacity to conduct a number of small-scale civilian missions concurrently with at least one large "substitution" mission, in which European staff temporarily replace the local government. European leaders aimed to be able to do so on short notice and in a non-benign environment. These missions were also to be sustainable for long periods of time (European Council, 2004b).

At the start, the EU laid out specific numerical targets, and, by 2004, member states had committed 5,761 police, 631 rule of law experts, 562 civilian administrators, and 4,988 civil protection staff for EU civilian missions (European Council, 2004a). These figures sound impressive, but because the commitments are nonbinding, the actual staff available for EU missions is much less, as recent difficulties in recruiting staff for specific missions have made clear. A recent study pointed out that, although there are some 1.6 million EU civilian personnel available, only 5,000 are pledged and some 2,000 deployed because of competing demands, often at home in Europe (Gya, 2009). As a result, the more recent Civilian Headline Goal 2010 lays out a process for assessing needs and improving recruitment success, eschewing the question of numerical targets.[1]

---

[1] See *EU Security and Defense: Core Documents 2007*, 2008. See also Howorth, 2007, pp. 124–133.

Basic Structures

In recent years, the EU has established certain institutional structures designed to facilitate its civilian capabilities (see Figure 2.1). As with ESDP military operations, civilian ESDP operations are run under the authority of the Political and Security Committee (PSC), which comprises the political directors of the foreign ministries of the member states and reports to the EU Council, which itself consists of the foreign ministers of the member states. With the implementation of the Lisbon Treaty, the EU will gain a High Representative for Common Foreign and Security Policy, who will represent both the Council and the Commission. This position will replace the current Secretary General/High Representative position, long held by Javier Solana, and unify the heretofore separate foreign policies of the supranational and the intergovernmental institutions of the EU. The High Representative for Common Foreign and Security Policy will also be closely involved in the strategic and operational management of ESDP, including the EU's civilian work.

As early as 2000, the Committee for the Civilian Aspects of Crisis Management (CIVCOM) was established to advise the PSC on the civilian dimension. CIVCOM brings together members of both the intergovernmental and supranational branches of the European Union and serves as a working group that makes recommendations and helps with strategy development on civilian issues. The main effect of CIVCOM has arguably been to ensure an appreciation of the importance of civilian work to postconflict and other crisis situations (Pfister, 2008, p. 194).

The most important institutional innovation in recent years, however, has been the establishment of a Civilian Planning and Conduct Capability (CPCC) in 2007. The CPCC serves as a headquarters for EU civilian missions, providing planning and operational support. It is staffed by some 70 experts in civilian operations and run by a civilian operations director. It also houses a 24-hour "Watch-Keeping" facility that tracks developments and provides situational analysis to the CPCC, PSC, and High Representative.

Figure 2.1
Basic Structures for Civilian ESDP

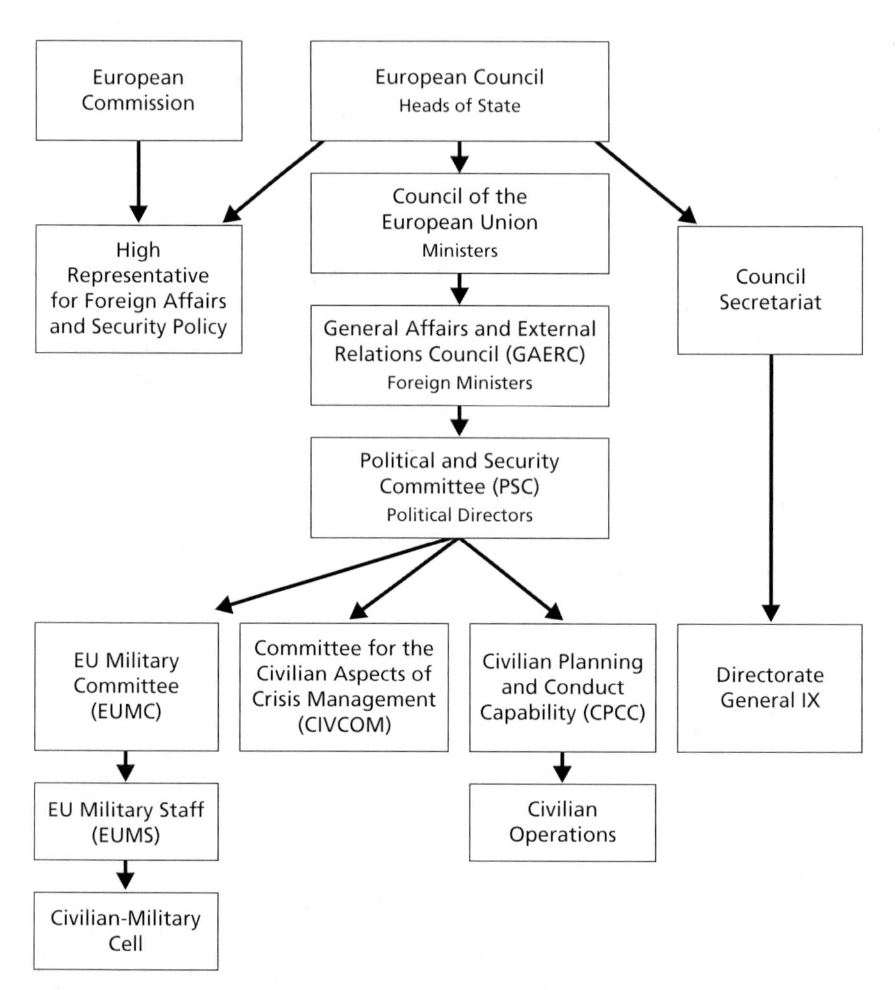

NOTE: Should the Lisbon Treaty fail, the High Representative position would remain the Secretary General/High Representative for Common Foreign and Security Policy (SG/HR), under the European Council only.

RAND *MG945-2.1*

The CPCC is separate from the civilian-military planning cell that has been established on the EU Military Staff (EUMS), but it is located in the same building, to facilitate coordination.[2] The establishment of the CPCC has in part replaced the role once played by the Council Secretariat's Directorate General IX, which, however, continues to serve as the point of origin for draft concepts of operations for the EU's civilian missions.[3]

Although, to date, joint civilian-military operations have not taken place, the EU aspires to deploy one, and there are ongoing efforts to foster closer cooperation. For example, the EU has been reorganizing the Council Secretariat to improve coordination between civilian and military ESDP missions, and Lt General David Leakey, who heads the EUMS from 2007 to 2010, has been working to improve civilian-military coordination.

General Record So Far

Since 2003, when the first civilian mission was launched in Bosnia, the EU has conducted civilian missions in 13 countries. This impressive figure can be misleading, however, given that, as Table 2.1 shows, the majority of these missions were small advisory missions. Nevertheless, there is an upward trend. As Figure 2.2 illustrates, the total number of EU civilians deployed abroad has grown from 715 in 2003 to nearly 2,800 in January 2009. A large part of this growth is the result of the very large EULEX Kosovo mission, which began in 2008. But it is nevertheless an indication that the ambitions and availability of EU civilian staff is on the rise, at least for some missions.

The EU's civilian aspirations differ in important ways from those of the United States. EU missions have, to date, been largely independent of military operations. Unlike Provincial Reconstruction Teams (PRTs), for example, they do not, in most cases, deploy in conflict

---

[2] On the CPCC, see Pfister, 2008, pp. 202–207.

[3] For useful organizational charts of ESDP, see International Crisis Group, 2005, pp. 6, 7, 17. See also Howorth, 2007, p. 69.

Figure 2.2
Growth of EU Civilian Deployments

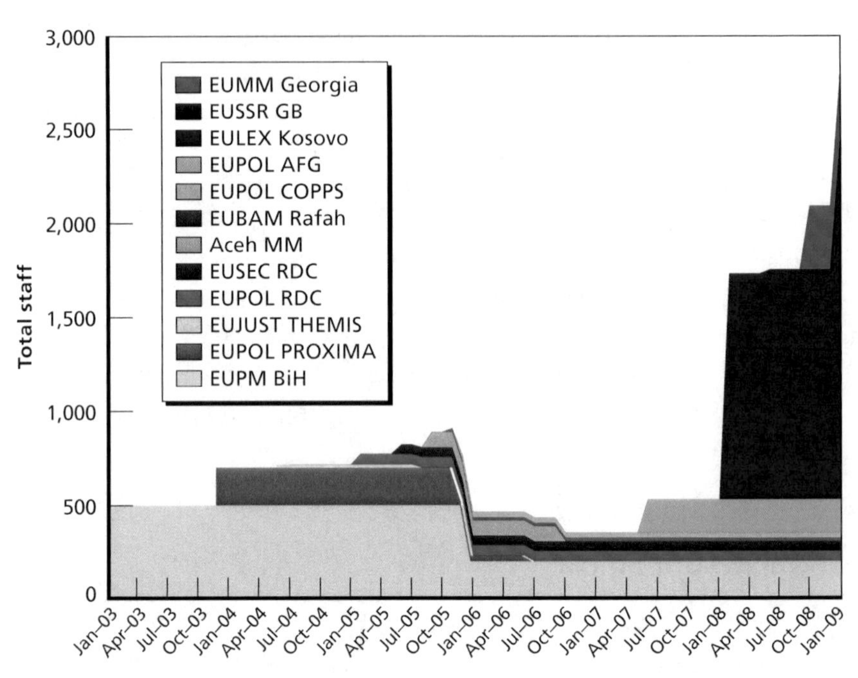

RAND *MG945-2.2*

zones, although some EU police are deployed to PRTs in Afghanistan. Most EU missions are advisory rather than executive in nature, although executive and replacement missions are not ruled out.

Police Missions

Police missions have been the most important EU civilian missions to date. The EU has deployed more police missions than any other kind of mission and has deployed more police advisers than any other type of personnel. This is in part a reflection of the fact that police missions tend to be more staff-intensive than monitoring or rule of law missions, but it is also a reflection of the EU's focus on policing and the rule of law, especially in the Balkans, where the absence of the rule of law has consequences for Europe's security.

Table 2.1
EU Civilian-Military Missions

Mission	Type	Dates	Country	Staff
EU Police Mission in Bosnia and Herzegovina (EUPM/BiH)	Police	1/2003–	Bosnia and Herzegovina	500
EU Police Mission in the Former Yugoslav Republic of Macedonia (EUPOL PROXIMA)	Police	12/2003–6/2006	Macedonia	200
EU Rule of Law Mission to Georgia (EUJUST THEMIS)	Rule of law	6/2004–7/2005	Georgia	27?
EU Police Mission in Kinshasa (EUPOL Kinshasa)	Police	2/2005–	Democratic Republic of Congo	58
EU Integrated Rule of Law Mission for Iraq (EUJUST LEX/Iraq)	Rule of law	6/2005–	Iraq	27
EU Advisory and Assistance Mission for Security Reform in the Democratic Republic of Congo (EUSEC RDC)	SSR	6/2005–	Democratic Republic of Congo	50
EU Civilian-Military Supporting Action to the African Union Mission in Darfur (EU Support to AMIS)	Police/ military	7/2005–12/2007	Sudan	50
Aceh Monitoring Mission (Aceh MM)	Monitoring	9/2005–9/2006	Indonesia	80
EU Border Assistance Mission in Rafah (EUBAM Rafah)	Monitoring	11/2005–	Palestinian Territories	20
Moldova and Ukraine Border Mission	Monitoring	11/2005–	Moldova and Ukraine	119
EU Police Advisory Team in the Former Yugoslav Republic of Macedonia (EUPAT FYROM)	Police	12/2005–	Macedonia	30
EU Police Mission for the Palestinian Territories (EUPOL COPPS)	Police	1/2006–	Palestinian Territories	32
EU Police Mission in Afghanistan (EUPOL Afghanistan)	Police	6/2007–	Afghanistan	177
EU Rule of Law Mission in Kosovo (EULEX Kosovo)	Combined	2/2008–	Kosovo	1,900
EU Mission for the Security Sector Reform in Guinea-Bissau (EUSSR GB)	SSR	6/2008–	Guinea-Bissau	21
EU Monitoring Mission in Georgia (EUMM Georgia)	Monitoring	10/2008–	Georgia	340

NOTE: Figures in the staff column indicate total authorized international staff, and thus in some cases may be higher than the actual deployment.

The main focus of the EU's police missions has been in the Balkans, where the EU has conducted training and advisory missions in Bosnia, Macedonia, and Kosovo. These missions have focused on confidence building, often between ethnic groups, helping local police develop interethnic police forces, fighting organized crime, and generally helping the host nation improve the quality and professionalism of its police forces.

Beyond the Balkans, the EU has also contributed to police work in the Democratic Republic of Congo (DRC), the Palestinian Territories, and Afghanistan (discussed in greater detail below). In the DRC, the EU focused mainly on general advice to the state's effort to establish a national police force while working to support the establishment of an integrated police unit. In the DRC, the EU worked alongside bilateral efforts of key European states, such as Belgium. In the Palestinian territories, the EU mission has provided advice and mentoring to the police of the Palestinian Authority, including training traffic police, helping modernize the Palestinian Authority's bomb squad, and mentoring the Palestinian Authority's own police training efforts.[4]

Rule of Law Missions

The EU has also sent rule of law missions to Georgia and Iraq, and the EULEX Kosovo mission has a major rule of law component. The EU mission in Georgia was set up after the 2003 "Rose Revolution" to support the democratization process. The mission worked to advise the Georgian government on judicial and prison reform, with the aim of preparing for stronger EU-Georgia ties.[5] It appears to have had little success in bringing about long-term reform, however, in part because of lack of interest from the Georgian political leadership. As one study concluded, the main value of the mission was that it was deployed relatively quickly at a critical moment in Georgia's political evolution, and thereby provided the EU with some extra political leverage (Helly,

---

[4] Smith, 2008; Berg, 2000; and EU Council Secretariat, 2009b. For general context, see Solana, 2008.

[5] "EU Sends Mission to Georgia to Reform Legal, Prison Systems," 2004; European Union Rule of Law Mission to Georgia, 2004; and Helly, 2006, pp. 87–102.

2006, p. 102). In retrospect, it is more accurately conceived as rule of law support than postconflict work.

The EUJUST LEX/Iraq mission is intended to improve the rule of law in Iraq by combining police training with training for judicial and prison officials (Bouilet, 2005). By early 2009, it had provided training to 2,000 Iraqis in these three areas ("EU Justice Experts Eye Move into Iraq," 2008; Gros-Verheyde, 2009). For both political and security reasons, however, the EU mission was run from Brussels, and most training took place in Jordan, other third countries, or in Europe itself. As the U.S. Institute of Peace has pointed out, "Though EULEX activities claim to give priority to indigenous community involvement, the majority of EUJEST-LEX activities take place inside the EU rather than in the field" (McFate, 2008). This clearly diminishes the potential impact of the EU's presence in Iraq, and is indicative of how security concerns have hampered the EU's ability to conduct aggressive civilian missions.

Some European officials argue that the EU mission in Iraq added value on account of the fact that it was not associated with the U.S.-led coalition of the willing.[6] This is debatable, but it can be said that the mission had a positive effect, though on a small scale when compared against the magnitude of the overall nation-building challenges Iraq faced.

Monitoring Missions

The EU has also deployed personnel to monitor ceasefires and borders, for example, at Rafah on the Gaza strip and on the Moldova-Ukraine border. In Aceh and Georgia, EU missions monitor ceasefires. The EU also monitored the Demobilization, Disarmament, and Reintegration process in Aceh, although the mission was relatively short-term and operated alongside missions from other international organizations. The fact that it took place in Asia has been touted, rather dubiously, as evidence of the EU's global reach (Pirozzi and Helly, 2005).

The recent EU mission in Georgia was launched to monitor the "six-point" ceasefire agreement between Georgia and Russia after Rus-

---

[6] Interviews with German officials, Berlin, May 2008.

sia's 2008 invasion. This mission was much larger than previous such missions. In fact, in this rare instance, there were too many volunteers, and the mission had to be expanded to accommodate the overabundance of European monitors.[7] The Georgia monitoring mission was nevertheless a case where recourse to the EU was advantageous for political and diplomatic reasons. The Organisation for Security and Co-operation in Europe (OSCE), which often plays this role, was not available because Russia is a member. Nor would NATO play a role, for obvious reasons. Stationing EU monitors in Georgia in large numbers also offered a deterrent effect against potential future Russian aggression.

Civil Administration Missions

Most of the missions in this category are smaller in scale and involve EU civilian or military staff serving in advisory roles in the national administration of third countries. For example, in the DRC, the EU established a mission for SSR. The majority of the staff were stationed in the Congolese Defense Ministry, in an effort to build capacity. This particular mission also helped provide the UN and EU with a clearer view of what was actually going on in the ministry and identified and then helped resolve problems with the ministry payroll system—a problem that had a major impact on the behavior of soldiers and situation on the ground.

In addition to an advisory and capacity building role, the EU is, in this area as others, also able to deploy civil administration for substitution missions, although it has not yet done so. As the experience in Kosovo indicates, however, the distinction between substitution and advice is easily blurred.

Security Sector Reform

The EU's recent deployment of a small SSR mission to Guinea-Bissau is also worth brief mention. The small mission is 15 strong and draws on the resources of six European countries. The main purposes are to advise on the restructuring of the Guinean armed forces and police and

---

[7] Interview with EU official, Brussels, November 10, 2009.

contribute to Interpol efforts to combat drug trafficking.[8] The mission parallels a similar EU effort in the DRC.

Civilian Response Teams

One final development that deserves mention are the Civilian Response Teams, which are intended to be small-size, multifunctional teams deployable on short notice in crisis situations. The teams would be modular, with different packages for different crisis situations. Their main use would be in the assessment phase of a crisis situation; supporting the deployment of an ESDP civilian mission; and providing temporary support for an EU Special Representative during a crisis (European Council, 2005).

---

[8] See Observatoire de l'Afrique, 2008; EU Council Secretariat, 2008.

CHAPTER THREE

EUPOL Afghanistan

So far, the EU has undertaken two civilian missions that operate as part of a larger nation-building effort in which NATO is a key actor: first, the EU Police Mission in Afghanistan (EUPOL Afghanistan); second, the EU Rule of Law Mission in Kosovo (EULEX Kosovo). Both of these missions have been relatively ambitious, although the Kosovo mission far more so, a fact that reflects Europe's long investment in stabilizing the Balkans. Each deserves a more in-depth treatment. This chapter thus examines the EUPOL Afghanistan mission, while Chapter Four looks at EULEX Kosovo.

The purpose of the EUPOL Afghanistan is to assist in establishing an effective Afghan civilian police force. It is a follow-on mission to the German Police Project Office (GPPO) that had been operating in Afghanistan since 2002 and was widely regarded as underperforming.[1] In 2007 and for much of 2008, EUPOL continued the poor performance of its German predecessor. There were deficiencies with the basic approach and size of the mission, as well as technical difficulties, for example, in procuring basic equipment. By mid-2009, however, the mission had overcome at least some of these problems and was ready to make a more significant contribution to the construction of an Afghan police force.

[1] See for example, Dempsey, 2006; DiManno, 2008; and Graw, 2007.

Background

Police reform has been one of the major challenges in the broader post-conflict stabilization and reconstruction effort in post-Taliban Afghanistan (Perito, 2009; Murray, 2007; International Crisis Group, 2007). During the Soviet and Taliban years, Afghanistan lacked an effective civilian police force. During the 1960s and 1970s, Afghanistan had a national civilian force that received assistance and training from both West and East Germany. With the Soviet invasion, however, the police increasingly took on a paramilitary role—with the support of the KGB—to fight the mujahideen. Deterioration of the police continued after the Soviet departure and subsequent factional fighting. When the Taliban took control of Kabul in 1996, they established a "Vice and Virtue Police," on the Saudi model, but did little to develop western-style civilian policing (Wilder, 2007, p. 3). Hence, by 2001, Afghanistan had been without functioning regular police for over 20 years (Murray, 2007, p. 109).

The international community therefore faced a number of hurdles when it came to police reform—these included both establishing institutional structures and training procedures and effectively training a large number of police officers. According to the arrangement laid out in the November 2001 Bonn Agreement, Germany was to take the lead role in this area. A GPPO was staffed with 40 officers, who began training senior Afghan police at the Kabul Police Academy. German efforts focused on providing long-term training by offering three-year and nine-month courses. Germany also worked on the reform of the Ministry of Interior and coordinated the international community's work and contributions in this area, but with little success. By 2005, the resources the German mission dedicated to police reform were significantly less than those dedicated by the United States, a fact that deprived the German mission of its legitimacy and political authority (Gross, 2009, p. 27).

The German mission was widely considered a failure. As a German diplomat later put it, "We had produced a cadet shop, but completely ignored the situation on the ground" (Zepelin, 2009). In part the problem was simply the mounting insurgency, which made Germany's civilian-centered approach problematic. The failure of the German mission was one reason the United States started the Combined Security Transition Command–Afghanistan (CSTC-A) mission in 2005. This, combined with the growing U.S. pressure for an increased European effort in Afghanistan in general, led Germany, which at the time held the rotating EU presidency, to seek to "Europeanize" its police operation. EUPOL was launched in June 2007 with a three-year mandate.[2]

By that point, the EU was already heavily involved in Afghan reconstruction through its financial contributions, which made it the second-largest donor after the United States. After the Bonn Agreement, the European Commission had drawn 4.93 million euros from the Rapid Reaction Mechanism to help support the political transition in Afghanistan. More importantly, between 2002 and 2006, the EU contributed approximately 3.5 billion euros in aid (Gross, n.d., p. 1). This puts commission funds to work for the same objectives as the EUPOL mission. Total EU and EU member state contributions to Afghan reconstruction from 2002 to 2010 topped $10 billion.[3] The main focus of EU development efforts has been rural development and alternative livelihoods, governance and rule of law, and health. Under the category of rule of law in particular, the EU is the largest contributor to the Law and Order Trust Fund, which funds the Afghan National Police's operating costs (EU Council Secretariat, 2009a).

Mandate, Structure, Staffing

Headquartered in Kabul, as of mid-2009, EUPOL had some 260 international staff from 19 EU member states, plus Canada, Croatia, New Zealand, and Norway, as well as 123 local staff (see Table 3.1). Of these staff, there were 166 police officers, 17 rule of law experts, and 75 civilian experts.[4] The total staff authorized was increased to 400 in November 2008 (Council of the European Union, 2008), although the

[2] European Council Joint Action 2007/369/CFSP, May 30, 2007.

[3] Eight billion euros. EU Council Secretariat, 2009c.

[4] Interview with EUPOL official, via telephone to Kabul, July 30, 2009.

Table 3.1
EUPOL Staffing by National Origin, as of March 11, 2009

Country	Staff
Czech Republic	3
Denmark	14
Estonia	1
Finland	7
France	6
Germany	44
Hungary	1
Italy	11
Latvia	2
Lithuania	3
Netherlands	11
Poland	3
Romania	5
Spain	12
Sweden	4
United Kingdom	15
Non-EU	
Canada	4
Croatia	2
Norway	6
New Zealand	3
International civilian experts	65
Total international personnel	222

SOURCE: EUPOL Afghanistan, Kabul.

mission was still far short of this figure in July 2009. In fact, at the time the authorized level was increased, the previous level still had not been met. As of mid-2009, it was still unclear when EUPOL would meet its fully authorized level.

Two-thirds of EUPOL staff are deployed in Kabul, with the remainder in 15 provinces. EUPOL staff in the provinces are deployed to PRTs led by the country of origin of the EU staff. For example, in the Italian PRT, there are EUPOL Carabinieri and other Italian legal staff.[5] EUPOL is under the political control and strategic direction of the Political and Security Committee. The EU SG/HR provides political guidance to the mission and its operations through the EU special representative in Kabul.[6]

EUPOL Afghanistan's stated objective is to help establish an effective civilian policing system under Afghan ownership. The mission is tasked in particular to[7]

- develop a police reform strategy working with the international community
- support the Government of Afghanistan's implementation of the strategy
- improve the coordination and the cohesion of international police reform efforts
- address linkages to the broader rule of law.

Like the German mission that preceded it, EUPOL's focus is largely on mentoring and advising senior level staff. Unlike the German mission, the EUPOL mission has, since the fall of 2008, begun to train staff in addition to mentoring and advising. It also serves to coordinate previously disparate European efforts, and it increasingly serves as a lead actor coordinating the overall international effort on police reform through the International Police Coordination Board (IPCB).[8]

The mission has three stated "strategic objectives" for both policing and the rule of law.[9] For policing, the objectives are:

- Improve police command and communication.
- Introduce intelligence-led policing to increase proactivity.
- Improve criminal investigations.

---

[5] Interview with EUPOL official, via telephone to Kabul, July 30, 2009.

[6] European Council Joint Action, 2007/369/CFSP, May 30, 2007, Article 9.

[7] European Council Joint Action, 2007/369/CFSP, May 30, 2007, Article 4.

[8] Interview with EUPOL official, via telephone to Kabul, July 30, 2009; interview with Dutch official, via telephone to The Hague, July 30, 2009.

[9] Interview with EUPOL official, via telephone to Kabul, July 30, 2009.

For the rule of law, the stated objectives are:

- Fight corruption.
- Improve cooperation between police and judiciary.
- Develop human rights and gender structure policy for the police and Ministry of Interior.

The mission does not have executive power, and it acts only in a coordinating and advisory role. Common costs, which include the headquarters, the in-country transport, and other operational costs, are funded by the Common Foreign and Security Policy (CFSP) budget. Funding for 2008–2009 was 65 million euros.[10]

Main Activities

EUPOL operates alongside CSTC-A. In contrast with the European mission, the U.S. operation focuses on training large numbers of local police (International Crisis Group, 2008). In general, CSTC-A training emphasizes paramilitary functions, in contrast to the EUPOL effort, which emphasizes civilian policing. While the EU views its mission as a component of the promotion of law and order, the United States has tended to view police as a security force with a counterinsurgency function (Gross, n.d., p. 28).

EUPOL also operates alongside a number of bilateral police efforts run by European member states through their PRTs. In general, European civilians deployed to PRTs on a bilateral basis focus on the bottom levels of the Afghan police force, while EUPOL officers focus on the higher-level staff. Because EUPOL staff are normally deployed to their national PRTs, police officers from the same country can be operating together on the same PRT, but under different flags. EUPOL also provides staff with specific expertise, such as in forensics or other high-skill areas.[11]

---

[10] European Council Decision, 2008/884/CFSP, November 21, 2008.

[11] Interview with EUPOL official, via telephone to Kabul, July 30, 2009.

The field is increasingly crowded, however, and, in addition to the U.S. and European bilateral missions, EUPOL will soon operate alongside a NATO mission. At NATO's April 2009 summit, the Alliance agreed to introduce a new training mission to complement CSTC-A and EUPOL. The NATO Training Mission–Afghanistan (NTM-A), which, unlike CSTC-A and EUPOL, will fall under the International Security Assistance Force (ISAF), will focus on both police and military training. Its purpose with regard to police is to bring greater coherence to the police efforts that are taking place outside EUPOL on a bilateral basis and help coordinate these efforts with EUPOL itself.[12]

Much of EUPOL's work takes place in Kabul, either through training of senior police officers or through direct training of Kabul's own police force. One of EUPOL's main focuses in 2009 has in fact been improving security and policing in Kabul. It has initiated the "Kabul City Police Project," which aims both to make the city more secure against terrorist attacks and reform the local police. On security, EUPOL trains police in profiling suicide bombers, inspecting vehicles, and reacting to threats. On the police reform side, EUPOL is working to build criminal investigation capability and generally improve community policing.[13]

Another major focus for 2009 has been support for the August 2009 presidential elections. To this end, EUPOL introduced a train-the-trainer program focused specifically on elections, and it produced 350 graduates, each of whom were to train 10 further officers, who would in turn train 10 officers for a total "cascaded" sum of 35,000 police with special election training. The final round was not complete by the presidential elections, but European officials emphasized that the skills will be valuable for future elections, including parliamentary elections scheduled for 2010. As part of this pre-election preparation, EUPOL has also given special training to 40 police generals on monitoring police conduct during the elections.[14]

---

[12] Interview with Dutch official, via telephone to The Hague, July 30, 2009.

[13] Interview with EUPOL official, via telephone to Kabul, July 30, 2009.

[14] Interview with EUPOL official, via telephone to Kabul, July 30, 2009.

Some sense of the nature of EUPOL's role in the provinces can be gained through a snapshot of Uruzgan province, where the Dutch, who lead the PRT, have implemented police projects on a bilateral basis as well as through EUPOL. Dutch work on policing in the province has focused on three areas: training, mentoring, and building infrastructure. The first two areas have been undertaken in cooperation with both ISAF and EUPOL, whereas infrastructure work has been undertaken on a bilateral basis:

- On training, the Dutch have built the first provincial police training center, which provides certification courses for Afghan National Police. The center is staffed by five Dutch members of EUPOL, as well as by Australian federal police. The funding for the center comes from the Dutch government, but the EUPOL commissioner in Kabul has undertaken political initiatives with the Ministry of the Interior to gain full Afghan support for the center.
- Mentoring work takes place in the field, where there are five mentoring teams.
- The infrastructure effort is focused on building police stations that are defensible against insurgent attacks. As of June 2009, the Dutch had opened two and were working on eight more. The infrastructure effort is purely bilateral.[15]

Assessment

As of mid-2009, the EUPOL mission had trained some 7,000 police (Zepelin, 2009). Although the EUPOL mission is responsible for more than training, the training figures offer the easiest benchmark of the EU's impact. Needless to say, the EU mission has faced several challenges and remains the object of much criticism.

---

[15] Interview with Dutch civilian official via telephone to Uruzgan, June 11, 2009.

Staff Shortages

First, the mission is chronically short on staff. Not only is the parallel U.S. training effort under CTSC-A several times larger, the EU has been unable to meet its own staffing goals. In response to U.S. and other pressures, the EU doubled the number of staff in 2008, but six months later had still failed to recruit and deploy more staff in significant numbers (International Crisis Group, 2008, p. 10).

The reasons for the EU's difficulty in deploying staff lie in both the nature of the Afghanistan mission and in the EU's own recruiting process. All staff on the civilian mission are volunteers, and the fact that the Afghanistan mission is considered more dangerous and is less convenient than missions in the Balkans has significantly hampered recruitment ("Wo sind die Ausbilder?" 2007, p. 116). In an attempt to make the Afghanistan jobs more attractive, the EU increased per diem rates, which were below those for some other missions, in March 2009.[16] Even if staff are willing, however, it is the responsibility of the member states and not the EU itself to release them from national duty. They are unwilling to do so, especially when it comes to experts that are in low supply domestically, such as police snipers or investigators with special forensic or other skills. In a country such as Germany, where the majority of the police are under the control of the state rather than federal government, the difficulty is compounded (as in the United States). The national government, in other words, may make a promise of staff in Brussels that it finds difficult to fulfill back home. Such problems explain, in large part, the slow deployment of European civilians to Afghanistan.

These staffing shortfalls are especially significant given that police training is an area on which most EU civilian missions have focused and that some U.S. observers believe the EU has large numbers of police ready for deployment abroad. Theoretically, the EU has large numbers of deployable professional police. In practice, however, member states have had repeated difficulty recruiting police for missions outside Europe in large numbers. Boosting staff has been difficult, despite the recognition that more police are needed. Difficulty recruiting staff is

[16] Interview with EU official, Pristina, April 2, 2009.

partly the result of the fact that large numbers of European police are currently deployed in the Balkans. Should that mission come to an end, it is likely that some of the police working there would be available for missions further afield, including Afghanistan.

Alleged Risk Aversion

Second, the EU mission has been criticized for its risk aversion. As Figure 3.1 shows, personnel are deployed in many parts of Afghanistan, but the majority are in Kabul. The remainder are deployed in provinces through PRTs, where conditions were relatively secure. Roughly two-thirds of the staff remain in Kabul.[17] The geographical reach of the

Figure 3.1
Deployment of EUPOL Personnel

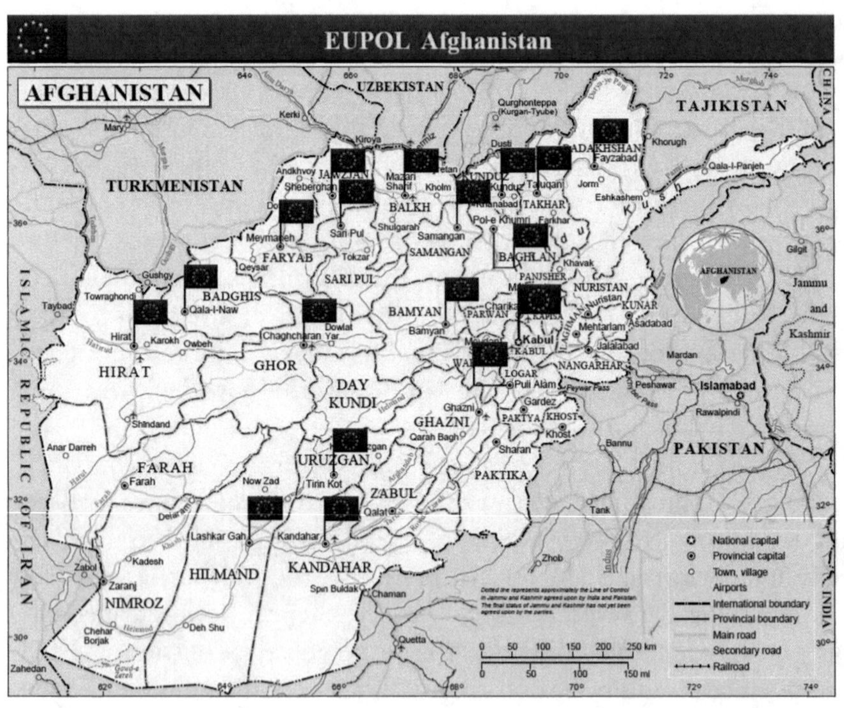

SOURCE: General Secretariat of the Council of the European Union based on original UN map of Afghanistan No. 3958 Rev. 5 produced in October 2005 by the UN Cartographic Section.
RAND *MG945-3.1*

[17] Interview with EUPOL official, via telephone to Kabul, July 30, 2009.

missions is thus limited. This is clearly an issue for the broader allied effort to build stability nationwide.

EU-NATO Impasse

Third, the EU deployment was hampered by the impasse over the EU-NATO working relationship. Turkey's refusal to allow the exchange of classified information with Cyprus has made it impossible to establish an EU-NATO security agreement, thus depriving the EUPOL mission of NATO protection. After several months of painstaking negotiations, bilateral agreements between the EU and the lead countries in each PRT were achieved. Because EUPOL police are normally deployed to their country of origin's PRT, this effectively meant organizing for countries to protect their own staff. Still the blockage of EU-NATO relations remains serious—in early 2009 there was, for example, no Blue Force Tracking for EUPOL staff.[18]

Logistical and Procurement Problems

Fourth, the mission has experienced many of the logistical and procurement problems that have plagued EU civilian missions elsewhere. Unlike military staff, who normally come with their own equipment, EUPOL civilian staff must be equipped. The EU rules governing procurement of a basic kit, however, are cumbersome and far too time-consuming for an operational deployment such as EUPOL. As a result, important equipment, such as armored vehicles and computers, were not in place when the mission began, prohibiting EUPOL staff from leaving base camp (Gross, 2009, p. 30). Eventually, exemptions from EU procurement regulations were given and now procurement is faster.[19]

Limited Size of Mission

Fifth, the overall small size of the EU mission, combined with the fact that it does not have an executive mandate, and thus is limited to advising and assisting the Afghan National Police, makes its success

---

[18] Interview with EU Official, Washington, D.C., December 5, 2008.

[19] Interview with Dutch official, via telephone to The Hague, July 30, 2009.

dependent on the will and receptiveness of the Afghan government. Without strong political and financial incentives, it can be difficult to see reforms implemented, especially when it comes to combating corruption (Gross, 2009, p. 34).

Summary

European advocates of the mission will argue in their defense that, despite its problems, the EU has managed to fulfill the function of training higher-level, higher-quality officers. In addition, they point to the EU's role as the provider of the secretariat of the IPCB, the body tasked to coordinate the international police reform effort, which, according to several sources, was defunct by 2005.[20]

The fundamental question is to what extent the problems the EUPOL mission has encountered are endemic to the EU and to what extent they arise out of the particular nature of the Afghan mission. In this regard, the EU's failure to meet its own staffing goals is particularly disappointing, especially when viewed in light of the fact that such shortfalls have been perennial. Without changes to recruitment efforts, these shortfalls are likely to continue. The limited size of the mission—even without counting the staffing shortfall—is not endemic to the EU, as the Kosovo mission suggests, although it does suggest that EU missions that take place outside Europe's periphery may be limited.

The logistical and procurement problems EUPOL Afghanistan has encountered, by contrast, should not be considered endemic. The EU recognizes these problems and has made efforts to make improvements in the future. Similarly, while criticisms of the EU strategy of training upper levels of police may be fair, choosing an inappropriate strategy is not irremediable or inextricably bound with the European approach. Moreover, a functioning Afghan police force will require officers at all levels, and if the EU trains the upper level, this is still a contribution, even if it is not as significant a contribution as the

---

[20] Interview with EU official, Brussels, November 13, 2008; interview with Dutch official, via telephone to The Hague, July 30, 2009; interview with EUPOL official, via telephone to Kabul, July 30, 2009.

United States might hope for given the EU's resources. The EU-NATO impasse, of course, cannot be blamed on the EU alone. In general, reforming Afghanistan's police is, inherently, extremely difficult, and criticisms of the EUPOL mission must always be viewed in this light.

Still, the shortages of staff and general lack of progress after seven years of effort (including the German effort) suggest that the future promise of EU civilian contributions, at least in hostile environments far from Europe, is limited.

The EU's effort in Kosovo, by contrast, suggests that the EU can make a serious contribution in smaller states, closer to western Europe. It is the subject of the next chapter.

CHAPTER FOUR

EULEX Kosovo

The EULEX Kosovo mission is the most ambitious civilian mission the EU has undertaken to date. Mandated by the European Council in February 2008 to strengthen the rule of law in Kosovo, EULEX is not only the largest such EU civilian mission, it is also the first integrated mission, with staff for police, rule of law, and customs and border patrol.[1] EULEX is also the first EU mission with executive power—that is, the power to intervene directly in Kosovo's affairs. EU officials on the ground, however, prefer to emphasize the mission's monitoring, mentoring, and advisory focus. Finally, EULEX has the added novelty of including over 70 U.S. staff, who are now successfully operating under EU authority.

Over the course of 2008, the mission managed a difficult deployment. EULEX faced problems that led some observers to wonder whether ESDP, which was born in the Balkans, might expire there.[2] In the end, however, the mission deployed to the controversial Kosovo Serb areas, although its status there was tenuous and its ability to carry out its mission restricted.

The problems the mission encountered are both political and technical in nature.[3] The technical issues can be resolved satisfactorily.

---

[1] European Council Joint Action, 2008/124/CFSP, February 4, 2008.

[2] Interview with EULEX official, Pristina, April 1, 2009; interview with ICO official, April 3, 2009.

[3] Some analysts have pointed out that hopes of keeping civilian-military work "technical" in postconflict environments are generally ill-founded. See, for example, Muehlmann, 2008.

The political problems, however, arise from divisions within the EU over Kosovo itself, and as such indicate some of the inherent limits not only of the EU's civilian capabilities, but also of ESDP in general.

Background

After NATO's 1999 air campaign to oust Slobodan Milosevic's forces from Kosovo, UNSCR 1244 gave NATO responsibility for most security tasks and a UN Mission, UNMIK, responsibility for administering Kosovo (Dobbins et al., 2003, pp. 111–128). Over the course of the next decade, the question of Kosovo's "final status" remained controversial. Serbia, supported by Russia, objected to plans for Kosovo independence. A plan for Kosovo independence with strong protections for Kosovo Serbs was negotiated by UN Special Envoy Martti Ahtisaari but failed to gain UN Security Council approval on account of Russian objections. As a result, Kosovo declared independence in February 2008.

Kosovo's declaration of independence without UN sanction created an ambiguous situation on the ground. According to the Ahtisaari plan, the EU was to deploy a civilian mission to replace UNMIK and help the government of Kosovo maintain and build the rule of law. The EU mission would be authorized by the now-sovereign government of Kosovo. As EULEX began to deploy in 2008, however, its status was still unclear. The EU sought to remain "status neutral"—meaning that its deployment would have no ramifications for Kosovo's status. This proved difficult, however.

The EU's focus on the rule of law dimension reflects, in part, prevailing EU philosophy of postconflict reconstruction. It also reflects, however, the more concrete interests of EU member states, who recognize Kosovo as a hotbed of international criminal networks and a center for trafficking, especially between the former Soviet Union and Western Europe. As one observer on the ground described it, Kosovo is a "cesspool" that collects contraband and holds it until an outlet is found in Europe.[4]

---

[4] Interview with western security expert, Pristina, April 1, 2009.

Mandate, Structure, Staffing

EULEX is an integrated police and rule of law mission whose purpose is to strengthen the rule of law in Kosovo. Unlike past EU civilian missions, the EULEX mission has executive authority for policing in some areas, hence the relatively large number of personnel. The mission mandate provided for several specific tasks, including[5]

- monitoring, mentoring, and advising Kosovo authorities
- reversing or annulling operational decisions of those authorities, when necessary to preserve the rule of law
- ensuring that judicial system is independent of political interference
- investigating or assisting in the investigation of war crimes, terrorism, organized crime, corruption, and other serious crimes
- improving coordination and cooperation of the Kosovo rule of law authorities
- fighting corruption.

EULEX funding was 205 million euros for the first 18 months.[6] Funding is for operations but does not pay basic staff salaries, which are paid by contributing states. Supplemental staff salaries, such as per diems, are paid through the EULEX budget. By 2009, the EU had spent, including through other mechanisms, over 2 billion euros on Kosovo's stabilization and reconstruction (Pond, 2008).

Strategic control of the mission falls to a Civilian Operational Commander located at the Civilian Planning and Conduct Capability (CPCC) in Brussels. He reports directly to the Political and Security Committee (PSC) and the EU's Secretary General/ High Representative for Common Foreign and Security Policy (SG/HR) and is expected to stay in contact with the EU Special Representative (EUSR) on the ground in Kosovo. The EUSR is double-hatted as the head of the International Civilian Office (ICO) but is not part of the EULEX

[5] European Council Joint Action 2008/124/CFSP, February 4, 2008, Article 7.

[6] European Council Joint Action 2008/124/CFSP, February 4, 2008.

mission and does not figure directly into the EULEX chain of command. Coordination, however, takes place both through the CPCC and the Head of Mission on the ground, who is responsible for operations, including liaison with other international organizations and the NATO Kosovo Force (KFOR) in particular. The mission headquarters were established in Pristina, with local offices elsewhere in Kosovo and a Brussels support element.

The majority of staff are seconded from national civil service, but many of the mission staff have prior experience in the Balkans, often because the staff were serving previously under UNMIK. Police Chief Rainier Kuehn, for example, has served in several capacities in the international effort in the Balkans since the mid 1990s.[7] Table 4.1 provides an overview of the balance between contracted and seconded staff, as well as the proportion of staff by national origin.

U.S. Participation

U.S. participation was thought to be convenient both because the U.S. traditionally has had good relations with the Albanian communities in the region, and, more practically, because U.S. staff were already on the ground as part of UNMIK. These staff were transferred to EULEX ("EU Eyes Deployment in Kosovo," 2008). These are civilian staff, of course, not military staff.

Main Activities

Policing

The police component is the largest of EULEX's three components. Figure 4.1 illustrates the proportion of staff dedicated to each component. It is organized around three police subcomponents. First, a strengthening component carries out mentoring, monitoring, and advisory work at the national and regional level. Second, a police executive component deals primarily with sensitive crimes, including war crimes, organized crime, corruption, and financial crimes. Third, a spe-

---

[7] Interview with Chief of EULEX Police, Pristina, April 3, 2009.

Table 4.1
EULEX Staffing by National Origin, as of April 2009

Country	Seconded	Contracted	Total
Austria	26	4	30
Belgium	39	5	44
Bulgaria	37	22	59
Cyprus	0	0	0
Czech Republic	23	1	24
Denmark	54	5	59
Estonia	6	2	8
Finland	58	18	76
France	181	13	194
Germany	132	9	141
Greece	30	7	37
Hungary	50	6	56
Ireland	9	7	16
Italy	176	25	201
Latvia	10	4	14
Lithuania	7	1	8
Luxembourg	2	0	2
Malta	0	1	1
Netherlands	30	4	34
Poland	120	12	132
Portugal	15	4	19
Romania	176	13	189
Slovakia	8	1	9
Slovenia	14	3	17
Spain	9	8	17
Sweden	82	6	88
United Kingdom	63	39	102
Non-EU			
Croatia	2	4	6
Norway	8	0	8

Table 4.1—Continued

Country	Seconded	Contracted	Total
Switzerland	7	0	7
Turkey	37	0	37
United States	76	0	76
Total international	1,487	224	1,711
Local staff			818
Grand total			2,529

SOURCE: EULEX Press Office, Pristina, April 2009.

Figure 4.1
EULEX Kosovo Police Breakdown Favors Special Police

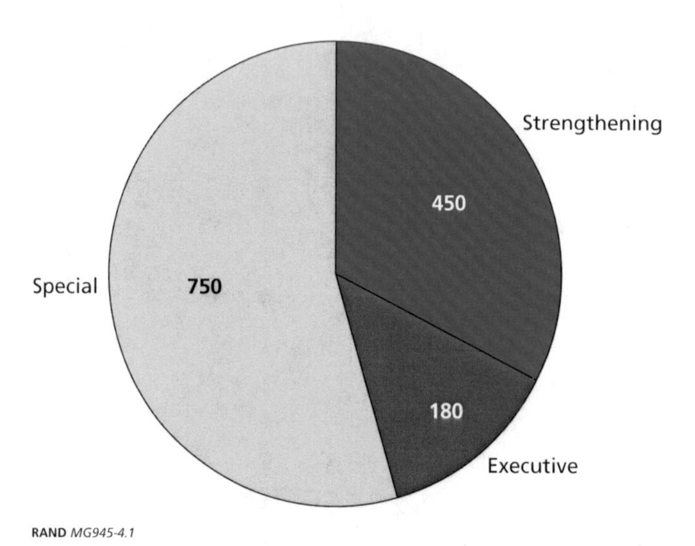

cial police acts as a gendarmerie force deployable in the event of civil disorder and available for close protection responsibility where needed.[8]

After experts questioned the preparedness of the Special Police Units stationed around Kosovo, units were repositioned in early 2009 to decrease response times, especially in the event of an emergency in

[8] Interview with EULEX official, Pristina, April 1, 2009.

the contentious North. As of 2009, the special units remain untested against large-scale civil unrest, but have successfully responded to smaller-scale demonstrations, as in Brdjani/Kroi i Vitakut.[9]

EULEX does not conduct police training, although it provides advice and mentoring to the Kosovo Police Academy.

Justice

The justice component comprises an international staff of approximately 250 judges and prosecutors and penal officers. It runs a missing persons office to deal with the 1,900 persons still missing from the conflict and the identification of the remains of some 400 persons. There are 70 judges and prosecutors from EU countries, the United States, Norway, Turkey, and Croatia. These judges have both mentoring and executive functions. This is a significant number in a country that has a total of 400 judges and prosecutors nationwide.[10]

One of the biggest challenges to the justice subcomponent was getting judges and prosecutors up to speed on Kosovo's multilayered legal code, which includes UNMIK code, Yugoslav code, Kosovo code, and even code of more ancient provenance. The question of which code to apply is political in nature and therefore contentious. Justice officials are forced to take into account the political ramifications of their decisions—rather than solely the legal. In addition, by its nature, the justice subcomponent faces an inherent challenge: On the one hand, it is part of an EU mission and reports to the head of mission; on the other hand, it must maintain judicial independence if it is to meet basic western standards. This can create difficulties if the rulings of individual EULEX judges run counter to the overall objectives of the mission.

The justice unit also faces challenges when prosecuting organized crime. Although combating organized crime is one of the more important tasks the EULEX mission must carry out—and arguably one of the key reasons Europe has invested heavily in the EULEX mission—

---

[9] Email exchange with EULEX official, Pristina, September 14, 2009.

[10] Interview with EULEX official, Pristina, April 2, 2009.

there has been some hesitancy to push all cases too far, for fear of destabilizing the political balance in Pristina.[11]

Customs

Customs is the smallest of the three subcomponents, but it is crucial to Kosovo's government finances, since tariff levies represent some 70 percent of government receipts (605 million euros in 2008).[12] EULEX operates in an advisory capacity at customs points around the country. There are four advisers in the Kosovo central administration and six mobile teams with three staff in each.

The most controversial of these have been the two customs gates along the northern border with Serbia. These gates were destroyed by Serb protestors a month after Kosovo gained independence. In consequence, between March 2008 and December 9, when the EULEX mission deployed, no customs were collected along the border, and a duty-free zone developed north of the Iber River. This created problems for Kosovo as well as Serbia. In Kosovo, oil and other goods from Serbia entered free of charge, often to be shipped back into Serbia itself, thereby avoiding excise taxes. Kosovo authorities set up checkpoints along the Iber to prevent goods from entering the south, but there were no obstructions on driving to the North to purchase goods tax-free. The system was a boon to criminality in the North, lining the pockets of some separatist Kosovo Serb leaders.

EULEX deployed successfully to the northern customs gates on December 9 but, as of April 2009, when full operational capability was announced, was still not collecting duties. Instead, EULEX assisted in the gradual establishment of a data collection system, involving photocopying the licenses of drivers passing though the gates. According to EULEX, this data collection effort has led to a significant drop in smuggling, with estimated revenue losses decreasing by over 80 percent.[13]

---

[11] Pond, 2008. Confirmed in various interviews in Pristina, April 1–3, 2009.

[12] Interview (1) with EULEX official, Pristina, April 2, 2009.

[13] Interview (1) with EULEX official, Pristina, April 2, 2009.

Assessment

In general, the EULEX mission must be viewed as the best evidence that the European Union is capable of making a real contribution to the civilian dimension of postconflict stabilization and reconstruction. This is true, notwithstanding the problems the mission has faced, as explained below.

Deployment Challenges

Russia, Serbia, and Kosovo Serbs persistently objected to the deployment of the EU mission on the grounds that it implied a de facto recognition of Kosovo independence by the EU ("UN Hands over to EU in Kosovo," 2008). On December 9, 2008, however, the EU successfully established a presence at the controversial North Mitrovica courthouse, previously the site of rioting against the international presence, and along the two disputed customs gates with Serbia.

As of April 2009, however, the EU presence in the North was far from complete. European customs officials have been deployed along the border with Serbia but only collect data—no duties. While European justices have established themselves at the controversial north Mitrovica courthouse that was the scene of rioting a year ago, in four months they have tried only one case. European police—an essential part of the mission—were not able to deploy to the disputed area and the chief of police of the European mission was himself forced to turn back during a recent visit after Serb protestors threw rocks at his vehicle.

Elsewhere in Kosovo, however, EULEX was operating successfully by the end of 2009.

Problems of a Political Nature

The main problems the EULEX mission has faced across Kosovo stem from internal European divisions over Kosovo's final status. European states that have not recognized Kosovo—notably Spain, Greece, Romania, Slovakia, and Cyprus—while supporting the mission, seek to ensure that EULEX does not take any action that favors Kosovo independence. EULEX officials have attempted to overcome the problem by emphasizing the technical nature of their mission and insisting that they will build institutions and the rule of law regardless of whether Kosovo is an independent state. But this can be an extremely difficult position to maintain in practice. Europe's division over Kosovo creates major problems when it comes to the legal system in particular. Kosovo has at least four competing legal codes—the Kosovo code, the UNMIK code, the Serb code, and one code of ancient provenance that predates Yugoslavia. When trying cases, EU justices must choose which legal code to imply—a choice that has clear political implications.

On a deeper level, the EU's problems in deploying to the Serb municipalities in the North can be seen as in part the result of the EU's reluctance to press Serbia to give up its financial and rhetorical support for the Serbs living in that area. The EU has significant influence over Serbia and could use that influence to press Serbia to prevail upon the Serbs in the North to allow the EU to function there fully, without fear of reprisals. The division over Kosovo, however, plus a lingering sense that the EU has not treated Serbia fairly in the past, prevents it from doing so.

The result is no small amount of frustration with Brussels in the EU mission, which must carry out its mandate within the confines of the EU's division.

Problems of a Technical Nature

The EULEX mission, however, has also faced technical problems. The first is staffing. While the EU has proven its ability to fill a large number of posts—far more than in Afghanistan—it has still fallen short of its target by several hundred staff. EULEX, like EUPOL Afghanistan, relies in large part on secondments from national civil services, in contrast to the United States, which relies largely on contractors. EU officials will argue that secondments are preferable, especially for mentoring at the higher levels of national administration, and that it is more difficult to find qualified contractors to serve as mentors to, for example, national police chiefs, though contractors may serve the purpose of doing police work themselves well, especially at the lower end.[14] The EU strategy for overcoming this problem is to increase per diems. Shortfalls in key specialized staff, however, may continue simply because European national governments are often reluctant to give them up, even when the staff are willing to deploy. Kosovo, of course, is not Afghanistan or Iraq, and the conditions, especially for Europeans who can easily return home on the weekends, are more conducive to the recruitment of civilian staff.

EULEX has also experienced significant problems with procurement and the handoff of essential logistics from UNMIK. As in other EU civilian missions, EULEX procurement takes place as if its officials were sitting in offices on the Avenue de Cortenberg in Brussels. As a result, every purchase requires tenders with three bids and must pass a number of other procurement regulations. The result has been unacceptably long delays in obtaining basic equipment for the mission, ranging form fax machines to desks to crucial police equipment.

The procurement issue is the more serious of the two technical difficulties the mission has faced, but it is also the easier to remedy. Indeed, awareness of the shortfalls is significant and will probably lead to necessary reforms such that EU civilian operations in the future will not be forced to cope with the added onus of procurement regulations written for very different circumstances.

Working Relationship with NATO

Although the Turkey-Cyprus problem precludes formal technical arrangements between EULEX and KFOR, the operational relationship is considered strong by both organizations.[15] The EU Planning Team deployed to Kosovo prior to the establishment of EULEX worked closely with NATO to develop arrangements for joint operations. These plans were then integrated into the respective organizations' operating procedures, though without any formal political agreement between the two organizations. Turkey is obviously aware of this arrangement but has chosen to signal that it will not object provided the two orga-

[14] Interview with European official in the ICO, Pristina, April 1, 2009.

[15] Interviews with EULEX and NATO officials, Pristina, March 31, April 1, and April 2, 2009.

nizations do not enter into formal arrangements. As a result, EULEX and NATO are in close contact at the lower levels and carry out regular training exercises, which are accorded the euphemism "technical discussions."

CHAPTER FIVE

Conclusions

Since its first mission in 2003, the EU has deployed civilians in several capacities and a variety of environments, ranging from benign to hostile. At the same time, the EU has continually reformed and worked to rationalize the relevant institutions in Brussels to improve the EU's civilian record. These missions, however, have been relatively small in scale, and so far have not, in general, had a major impact on security challenges of significance to the United States. The notable exception is the integrated mission in Kosovo, and perhaps the EU Police Mission in Bosnia-Herzegovina.

EU missions have faced several hurdles, some of which are inherent in the consensual nature of the EU's Common Foreign and Security Policy, some of which are of a technical nature and likely to be resolved, and some of which are beyond the EU's control and derive from the nature of civilian work itself. If the EU is to improve and expand its record in the future, the single most important task will be to improve its record for staffing its missions.

Nevertheless, despite the relatively small scale of most EU missions to date, the EU is poised to make a more significant contribution to allied civilian needs in the future, especially if the future EU missions follow the Kosovo model.

Overcoming the EU's Staffing Problems

On paper, the EU's potential when it comes to civilian operations looks impressive. Unlike the United States, several EU member states have national police that, in theory, should be easier to deploy to conflict zones than U.S. police. In practice, however, the EU has had difficulty deploying police in numbers that reflect its potential. In mid-2009, fewer than 1,800 of the more than 5,000 police staff committed by member states were deployed. This was not for lack of demand—the EU missions in both Kosovo and Afghanistan lacked staff. Indeed, the number deployed abroad today—especially given the size of the Kosovo mission—may represent the EU's real maximum level of deployment, unless a major effort on this front is undertaken in the future. However, the EU has generally had more success deploying civilian police than the United States, which has frequently been forced to rely on military police in postconflict situations.

In part, the EU's shortfalls are the result of the current high level of deployments, which have grown significantly since the end of 2007, with the deployment of EULEX Kosovo. Still, the fact remains that the EU has deployed far fewer staff than have been pledged by member states. The gap results in part from the collective nature of the process by which the EU recruits staff from member states: Member states can agree to EU missions collectively without having to commit resources individually, and when the bill comes, there is often not enough money on the table to cover it. It also results from the fact that member states can be reluctant to part with civilian experts that are sorely needed back home. In addition, from the perspective of the individual civilian, such deployments mean separation from family and living in uncomfortable and often dangerous environments. Sometimes, they are even detrimental for career advancement. It is perhaps for these reasons that many of the EU civilians who do deploy are former military.

One proposed solution for overcoming the staffing problems is to establish national contingents within the EU missions—in other words, change the procedure by which missions are approved, such that states would commit resources as part of the process. This has been resisted, however, on the grounds that it is not in the spirit of EU collective action. A second solution would be to increase the use of civilian contractors. But this could prove expensive, and many EU officials believe that the EU's added value is in quality and that high-quality civilian contractors are hard to come by. A third possible solution is to increase political pressure on states that do not appear to be living up to their commitments.[1] This would require a leadership role from one or more of the EU's larger states, but these states—Germany, for example—are among the culpable.

A fourth possibility is to increase EU funding for missions, which would reduce the financial burden on states that do send staff, especially when it comes to less prosperous states of the EU, some of which have national police. A fifth possibility is to further increase the per diem rates that are paid to civilians deploying in more dangerous areas. However, this has so far yielded only marginal gains in staffing the EUPOL mission in Afghanistan.

Sixth, and more dramatically, the European Union could resolve some of its staffing problems by developing a large pool of readily deployable civilians at the European level. This would obviate the need to go through the national states when a mission was agreed on, thereby potentially increasing both the speed and size of deployments.

The United States is in the process of establishing such a corps, under the direction of the Coordinator for Reconstruction and Stabilization (S/CRS). The U.S. corps will constitute three tiers, the first comprising 250 staff ready for deployment on 72 hours notice. The second and third would comprise 2,000 staff each, but with slower deployment times.[2]

The European Union should do the same, identifying, training, and making advance arrangements for the deployment of a select group of civilian experts. An even more ambitious model would be for the EU to establish a standing, small or even medium-sized body of civilian experts, who would be stationed and train together on a permanent basis, without responsibilities to their national governments. This would in many ways be the ideal solution to the EU's staffing problems, if it were politically and financially feasible. The only downside would be the possibility that such a corps might never be used, on account of the EU's own internal divisions when it comes to Common Foreign

---

1 Interview with Dutch Official, via telephone to The Hague, July 30, 2009.

2 Ambassador John Herbst, talk at the RAND Corporation, Washington, D.C., September 25, 2009.

and Security Policy. This possibility might seem remote, but the failure to deploy the EU battlegroups suggests that it is not unthinkable.

EU Added Value on Civilian Missions: Generic Considerations

What is the EU's value added when it comes to civilian operations? From the generic perspective, the main benefit of the EU is that, when compared with the bilateral approach, the EU approach allows for a greater aggregation of resources. In cases where needs are large or no single state is inclined to send staff in large numbers, this power of aggregation is significant, and it increases the chances that a civilian mission will be deployed. As it develops, the EU should also establish mechanisms for aggregating knowledge, experience, and lessons learned. This would also provide added value. Given that the CPCC's focus should be operations, this may require establishing a separate body for post-operational review. This body could be located on the high representative's staff, which should grow with the ratification of the Lisbon Treaty.

Another generic argument in favor of encouraging investment in the EU capabilities is that the EU may, in some situations, offer an attractive alternative when NATO or the United Nations are not viable for political reasons. This was the case in Georgia, for example, where NATO or UN monitors would not have been viable. The possibility of other similar situations in the future, especially in the Middle East, suggests that further investment in the EU option is worthwhile. This, of course, does not preclude continued investment in the capabilities of the UN or NATO.

A third possible benefit often cited is that the EU brings together economic, political, and military power in a unique combination and unprecedented scale, and can therefore coordinate financial aid and crisis response. Although accurate in theory—as it is for the U.S. government—this advantage has proven limited in practice. Although the Commission does focus on many of the areas where ESDP missions are running, this is not because the EU is running civilian missions in these areas, but simply because the areas are of interest to the European Union. If the missions were run by NATO, in other words, the European Commission would still focus on them. By establishing a single figure who speaks for the Commission and Council on foreign policy, however, the Lisbon Treaty may help to boost coordination, provided that the individual chosen is able to be effective in such a far-reaching role.

The EU's Added Value: Considerations for the United States

From the U.S. perspective, the added value of the EU's civilian-military work is that it will increase the chances that European states will contribute to such efforts outside Europe. Although there is an argument to be made for building civilian capabilities within NATO, this is a political non-starter in most European capitals, and it is clearly better to have a solid EU capability that can be deployed alongside NATO than to have little or no capability whatsoever. To be sure, so far the EU has only managed to deploy significant numbers of civilian experts in Kosovo, and unless the staffing issues discussed above are resolved, it may not be able to move further than this.

This is a reason for managed expectations when it comes to the EU's capabilities, but not a reason to oppose the development of those capabilities. Most major EU civilian missions have taken place alongside NATO, as a contribution to broader allied aims. This trend is apt to continue in the future. The U.S. thus should be willing to offer political and, where feasible, logistical and other support to EU civilian missions.

The NATO-EU Impasse

For the United States or Europe to obtain the full benefit of the EU's civilian capabilities, the operational impasse between NATO and the EU will have to be resolved. Although the experience in Kosovo shows that willing commanders on the ground can, given sufficient time, develop systems to work around the problem, the fact remains that the impasse is unnecessary, creates extra work, and saps morale and goodwill between the two organizations. Although Turkey will cite a host of grievances it feels the EU must resolve before it will agree to joint missions, it is possible that a broader improvement in Turkey's EU prospects may in fact be necessary. In either case, the problem remains and must be resolved.

Military Versus Civilian?

Finally, given the fact that there are those in the United States and elsewhere who would still prefer to see the EU remain a purely civilian power, it is important to note that the development of EU civilian capabilities should not become a substitute for the development of European military capabilities. While focusing on EU civilian capabilities may be an attractive option financially and politically within Europe, European leaders must be careful to avoid misleading European publics into the belief that these capabilities obviate the need for continued investments in traditional military training and hardware. From the U.S. perspective, while there may be little the United States can do to stem the decline of EU defense capabilities, it can at least avoid encouraging a solely civilian ESDP.

Indeed, if ESDP is to improve the security of European states, and offer European leaders the flexibility that is its root justification, it will be crucial to not only develop civilian capabilities, but also to continue and accelerate the long-standing effort to reform European national militaries so that they can deploy to conduct missions across a new spectrum of tasks. Europe should aim, in particular, to develop the capability to offer protection to civilian-military missions without the need to rely heavily on NATO for protection—as they effectively have in both Kosovo and Afghanistan.

References

Barton, Frederick, et al., *Civil-Military Relations, Fostering Development, and Expanding Civilian Capacity*, Washington D.C.: Center for Strategic and International Studies, forthcoming.

Berg, Raffi, "Rebuilding the Palestinian Police," BBC, November 30, 2000.

Bouilet, Alexandrine, "L'Europe en soutien de la loi en Irak," *Le Figaro* (France), February 22, 2005.

Cohen, Craig, and Noam Unger, *Surveying the Civilian Reform Landscape*, Washington, D.C.: Center for New American Security, 2008.

Council of the European Union, "2903rd Meeting of the Council: General Affairs and External Relations," press release, No. 15396/08, 2008.

Dempsey, Judy, "Training of Afghan Police Is Criticized, Germany Bears Brunt of NATO Pressure," *International Herald Tribune*, November 16, 2006, p. 4.

DiManno, Rosie, "Who Needs the Germans?" *Toronto Star*, May 30, 2008, p. AA01.

Dobbins, James, John G. McGinn, Keith Crane, Seth G. Jones, Rollie Lal, Andrew Rathmell, Rachel M. Swanger, and Anga R. Timilsina, *America's Role in Nation-Building: From Germany to Iraq*, Santa Monica, Calif.: RAND Corporation, MR-1753-RC, 2003. As of November 13, 2009: http://www.rand.org/pubs/monograph_reports/MR1753/

EU Council Secretariat, "Background on EU Mission in Support of Security Sector Reform in the Republic of Guinea-Bissau," February 2008.

———, "EU Engagement in Afghanistan," fact sheet, March 2009a. As of November 13, 2009: http://www.consilium.europa.eu/uedocs/cmsUpload/090330-EU_engagement_Afghanistan.pdf

———, "European Union Police Mission for the Palestinian Territories (EUPOL COPPS)," fact sheet, March 2009b. As of November 13, 2009:
http://www.consilium.europa.eu/uedocs/cms_data/docs/missionPress/files/090325%20FACTSHEET%20EUPOL%20COPPS%20-%20version%2012_EN.pdf

———, "EU Police Mission in Afghanistan (EUPOL Afghanistan)," fact sheet, July 2009c. As of November 13, 2009:
http://www.consilium.europa.eu/uedocs/cms_data/docs/missionPress/files/090708%20FACTSHEET%20EUPOL%20Afghanistan%20-%20version%2015_EN.pdf

"EU Eyes Deployment in Kosovo," *Jane's Intelligence Digest*, November 10, 2008.

"EU Justice Experts Eye Move into Iraq," Agence France Press, November 26, 2008.

EU Security and Defense: Core Documents 2007, compiled by Catherine Glière, Paris: European Union Institute for Security Studies, 2008.

"EU Sends Mission to Georgia to Reform Legal, Prison Systems," Agence France Press, July 16, 2004.

European Council, *Conclusions of the Presidency*, Santa Maria da Feira, June 19–20, 2000. As of November 13, 2009:
http://www.europarl.europa.eu/summits/fei1_en.htm

———, *Civilian Capabilities Commitment Conference: Ministerial Declaration*, Brussels, November 22, 2004a.

———, *Document 15760/04*, Brussels, December 2004b.

———, *Document 10462/05*, Brussels, June 23, 2005.

European Council Joint Action 2007/369/CFSP, May 30, 2007.

European Council Joint Action, 2008/124/CFSP, February 4, 2008.

European Council Decision, 2008/884/CFSP, November 21, 2008.

European Union Rule of Law Mission to Georgia, "Facts on EUJUST Themis," fact sheet, October 26, 2004. As of November 13, 2009:
http://www.consilium.europa.eu/uedocs/cmsUpload/Factsheet%20THEMIS%20041026.pdf

Giegerich, Bastian, *European Military Crisis Management: Connecting Ambition and Reality*, Adelphi Paper No. 397, London: Routledge, 2008.

Graw, Ansgar, "Polizisten in Afghanistan sind nicht viel mehr als Kanonenfutter: Streit um Ausbildung der Sicherheitskräfte," *Die Welt* (Germany), October 22, 2007, p. 4.

Gros-Verheyde, Nicolas, "Interview with Stephen White, Head of EUJUST Lex Mission for Iraq," *Europolitics,* January 21, 2009.

Gross, Eva, "Security Sector Reform in Afghanistan: The EU's Contribution," European Union Institute for Security Studies, Occasional Paper, No. 78, April 2009.

———, *The EU in Afghanistan—Growing Engagement in Turbulent Times*, Berlin, Germany: Heinrich Böll Foundation, no date.

Gya, Giji, "Tapping the Human Dimension: Civilian Capabilities in ESDP," *European Security Review*, No. 43, International Security Information Service–Europe, March 2009.

Helly, Damien, "EUJUST Themis in Georgia: An Ambitious Bet on Rule of Law," in Agnieszka Nowak, ed., *Civilian Crisis Management: The EU Way*, Paris: European Union Institute for Security Studies, Chaillot Paper, No. 90, June 2006.

Howorth, Jolywon, *Security and Defence Policy in the European Union*, Houndmills, UK: Palgrave MacMillan, 2007.

Hunter, Robert E., Edward Gnehm, and George Joulwan, *Integrating Instruments of Power and Influence: Lessons Learned and Best Practices*, Santa Monica, Calif.: RAND Corporation, CF-251-NDF/KAF/RF/SRF, 2008. As of November 13, 2009:
http://www.rand.org/pubs/conf_proceedings/CF251/

International Crisis Group, "EU Crisis Response Capability Revisited," *Crisis Group Europe Report*, No. 160, January 17, 2005.

———, "Reforming Afghanistan's Police," *Crisis Group Asia Report*, No. 138, August 30, 2007.

———, "Policing in Afghanistan: Still Searching for a Strategy," *Asia Briefing*, No. 85, December 2008.

Interview with Dutch official via telephone, The Hague, July 30, 2009.

Interview with Dutch civilian official via telephone, Uruzgan, June 11, 2009.

Interview with EU official, Washington, D.C., December 5, 2008.

Interview with EU official, Brussels, November 13, 2008.

Interviews with EULEX and NATO officials, Pristina, March 31–April 2, 2009.

Interview with EUPOL official via telephone, Kabul, July 30, 2009.

Interview with European official in the ICO, Pristina, April 1, 2009.

Interview with Head of EULEX Police, Pristina, April 3, 2009.

Interview with western security expert, Pristina, April 1, 2009.

McFate, Sean, *Securing the Future: A Primer on Security Sector Reform in Conflict Countries*, Washington D.C.: U.S. Institute of Peace, Special Report, No. 209, September 2008.

Menon, Anand, "Empowering Paradise? The ESDP at Ten," *International Affairs*, Vol. 85, No. 2, 2009, pp. 227–246.

Muehlmann, Thomas, "Police Restructuring in Bosnia-Herzegovina: Problems of Internationally-Led Security Sector Reform," *Journal of Intervention and Statebuilding*, Vol. 2, No. 1, March 2008, pp. 1–22.

Murray, Tonia, "Police-Building in Afghanistan: A Case Study of Civil Security Reform," *International Peacekeeping*, Vol. 14, No. 1, February 2007.

Nowak, Agnieszka, ed., *Civilian Crisis Management: The EU Way*, Paris: European Union Institute for Security Studies, Chaillot Paper, No. 90, June 2006.

Observatoire de l'Afrique, *Security Sector Reform (SSR) in Guinea-Bissau*, Africa Briefing Report, Egmont Palace, Brussels, February 2008.

Perito, Robert, *Afghanistan's Police: The Weak Link in Security Sector Reform*, Washington, D.C.: U.S. Institute of Peace, Special Report, No. 227, August 2009.

Pfister, Stéphane, "La Gestion Civile Des Crises: Un Outil Politico-Strategique Au Service De L'union Européenne," unpublished Ph.D. thesis, University of Geneva, December 2008.

Pirozzi, Nicoletta, and Damien Helly, "Aceh Monitoring Mission: A New Challenge for ESDP," *European Security Review*, No. 27, International Security Information Service–Europe, October 2005.

Pond, Elizabeth, "The EU's Test in Kosovo," *The Washington Quarterly*, Vol. 31, No. 4, Autumn 2008, pp. 97–111.

Smith, Colin, "High Noon for PA Civil Police," *Haaretz* (Israel), June 19, 2008.

Solana, Javier, "Erst das Fundament, dann das Dach," Sueddeutsche (Web site, in German), June 24, 2008. As of November 13, 2009:
http://www.sueddeutsche.de/politik/654/446390/text/

Thomson, Patricia, and Daniel P. Serwer, "Civilians Can Win the Peace," *USIP Briefing*, Washington, D.C.: U.S. Institute of Peace, 2007.

"UN Hands over to EU in Kosovo," *Jane's Country Risk Daily Report*, August 20, 2008.

Wilder, Andrew, "Cops or Robbers? The Struggle to Reform the Afghan National Police," Afghanistan Research Evaluation Union, Issue Paper Series, July 2007.

"Wo sind die Ausbilder?" [Where are the teachers?] *Der Spiegel* (Germany), December 17, 2007.

Zepelin, Joachim, "Zweiter Bildungsweg," *Financial Times Deutschland* (Germany), July 27, 2009. As of November 13, 2009:
http://www.ftd.de/politik/international/:Agenda-Zweiter-Bildungsweg/545075.html